203394

D1647020

WITHDRAWN

LIFE IN
THE MINE

ANTHONY BURTON

WITHDRAWN

IMPORTANT DATES

c.2300 BC Flint mining begins at Grimes Graves, Norfolk.

c.2000 BC Mining for copper and tin begins.

c.500 BC First iron ore mines opened.

AD 43 Romans settle in Britain and introduce new mining technology.

1228 Earliest record of coal being sent from Tyneside to London.

1294 370lbs (170kg) of silver ore sent from Devon to King Edward I.

1556 Agricola's *De Re Metallica* provides the first detailed illustrations of mining technology.

1686 First account of the use of a furnace for mine ventilation.

1705 First recorded colliery explosion at Gateshead: 30 miners died.

1712 Thomas Newcomen's steam engine installed at a coalmine in Dudley, Staffordshire.

1769 James Watt takes out a patent for his steam engine.

1776 John Curr develops a system for moving trucks underground on iron rails.

1812 Opening of the world's first successful steam railway at Middleton Colliery, Leeds.

1815 Sir Humphry Davy designs his safety lamp.

1863 Invention of mechanical coal cutter.

1919 Cornwall's worst disaster at Levant Mine.

1947 Coal mines nationalized.

1985 The year-long miners' strike against pit closures ends: 25 pits are closed immediately.

1991 Wheal Geevor, the last deep tin mine in Cornwall, closes.

2008 Tower Colliery, the last deep coal mine in South Wales, closes.

AN ANCIENT INDUSTRY

Men have been burrowing deep under the ground in Britain for more than 4,000 years. It all began in the New Stone Age when tools such as axes and knives were made out of flint. Some flint could be picked up at the surface, but the very best lay under a thick layer of chalk. Around 2300 BC, miners began digging pits as deep as 40 feet (12m) in an area now known as Grimes Graves in Norfolk. From the bottom of a pit they dug narrow tunnels through the chalk and began excavating the flint, using antlers as pickaxes and animals' shoulder blades as shovels. The only illumination consisted of little bowls made from chalk that would have been filled with animal oil and lit to produce a flickering flame. Altogether 300–400 pits were sunk.

The Stone Age gave way to the Bronze Age (*c*.2300–700 BC) and then the Iron Age (*c*.700 BC–AD 43). Now miners went in search of the raw materials, the ores that could be smelted to make the different metals. Copper and tin that were used to make bronze were mostly found in south-west England, and merchants came to Cornwall from as far away as the eastern Mediterranean to trade in the valuable metals. Ancient copper mines have also been identified at Great Ormes Head in North Wales. Here the miners had to cut through solid rock, and they used a technique known as fire setting. A fire was lit against the rock face and allowed to burn fiercely. When the rock became red hot, water was dashed against it, splintering the stone. A similar technique was used at iron mines such as those at Clearwell Caves in the Forest of Dean, Gloucestershire.

Exploring the complex of connecting tunnels and wider excavated areas in the Grimes Graves Neolithic flint mines in Norfolk.

T'OWD MAN

The strange name 'T'Owd Man' is Derbyshire dialect for 'The Old Man', and was used by the lead miners of that area to describe their predecessors who had started the industry. It almost certainly started with the Romans who really began the modern age of mining in Britain. They continued to work the older mines and also opened up new areas of exploration, such as the gold mines at Dolaucothi in Mid Wales. Here they introduced a whole new range of technologies. They began with a process called 'hushing'. This involved creating a reservoir of water, which could be released to cascade down a hillside, sweeping away the thin surface soil to reveal the veins of ore underneath. To get the water to the top of the hill in the first place they had to construct aqueducts, running many miles to the nearest river. They also brought in a new machine, the waterwheel, to power pumps to drain water from the workings. The men who worked there enjoyed the luxury of a bathhouse – something British coal miners did not get on a regular basis until the second half of the 20th century.

STAKING A CLAIM

By the Middle Ages, mining for metals had become well established with its own laws and rules. The lead mining region of Derbyshire was owned by the Crown and known as the King's Field of High Peak, governed by rules dating from 'a time whereof the memory of man runneth not to the contrary'. Anyone could lay claim by setting up a scowse, a simple hand-operated winch used for winding material up from the mine, over the spot. When the miner had enough ore to fill an official dish, which held just under 500 cubic inches (equivalent to 8 litres), he took it to the Barmaster; if the Barmaster approved, the miner could make his claim. He was allowed a length of land extending one 'meer' (about 30 yards/27m) either side of

◀ Medieval miners with their implements and tools, carved on a tombstone at Lasswade Colliery, Midlothian.

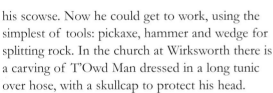

A Illustrations from Agricola's book *De Re Metallica* (1556), showing miners with a variety of devices for raising and lowering machinery in a pit (left), and different ways of entering a mine (right).

his scowse. Now he could get to work, using the simplest of tools: pickaxe, hammer and wedge for splitting rock. In the church at Wirksworth there is a carving of T'Owd Man dressed in a long tunic over hose, with a skullcap to protect his head.

When the ore was brought to the surface it had to be dressed to separate the valuable ore from the waste. This job usually went to women, who broke up the ore with a 6lb (2.7kg) hammer. Once most of the waste was removed, the pulverized ore was put in a jigger, a perforated box that was shaken in a tub of water. The heavy lead-bearing material sank to the bottom.

Mining was still largely undertaken by individuals who worked their own claims, rather than by organized companies, during this period.

A MINER'S HOME

Daniel Defoe, author of *Robinson Crusoe*, visited a lead miner's home in the 1720s: 'There was a large hollow cave, which the poor people by two curtains hang'd across, had parted into three rooms. On one side was the chimney, and the man, or perhaps his father, being miners, had found means to work a shaft or funnel through the rock to carry the smoke to the top … the habitation was poor, 'tis true, but things within did not look so like misery as I expected.'

DEEP PITS

As mines went ever deeper, so the miners were faced with new problems – particularly when they reached the water table. Getting rid of the water was expensive, and the mining industry could no longer rely just on individuals hacking away underground: capital was needed. In the 17th century, Sir George Bruce owned a coalmine at Culross in Scotland that had become flooded. He installed an 'Egyptian wheel' that consisted of an endless chain fitted with 36 buckets and driven by three horses. As the chain turned, the buckets dipped into a well at the bottom of the pit, then were carried back up and emptied into a trough at the surface. The mine extended for a mile (1.6km) under the Firth of Forth and became so famous that King James I came on a visit. He descended the shaft on the shoreline and explored the underground passages. He had not been told that he would come up by a different shaft that emerged on an island. When he came out and found himself surrounded by water he feared the worst, and cried out, 'Treason, treason!' He was eventually pacified.

A LONG WAY DOWN

Deep mines also provided new problems of access, often involving long climbs up and down the shaft. Nine-year-old William Crago, continuing

▼ Shot firing: setting off explosive charges at the coalface.

⌃ Stemming: using a metal rod to ram home a charge ready for blasting. The main difference between this modern photo and the earlier system is that the charge is ignited electrically, instead of by a fuse.

Iron rails or plates were introduced underground long before the age of steam railways. They made moving material far, far easier as this anonymous poem of the period in the Tyneside dialect affirms:

> God bless the man wi' peace and plenty
> That first invented metal plates,
> Draw out his years to five and twenty
> Then slide him through the heavenly gates.

the tradition of generations of sons following fathers down the mines that carried on right through to the 20th century, described his first working day in a 19th-century Cornish copper mine. He had to climb down a vertical ladder into the darkness, lit only by a candle stuck on his hat. 'We at last stepped onto the footway,' he said. 'Father first and I following him very carefully we descended the first 480 feet [145m]. It was almost climbing down the side of the house and as we slowly went along, ever and anon came Father's warning voice, "Hold tight your hands, my son."' After that it became slightly easier, as the ladders were sloping instead of vertical. Father and son finally reached the bottom of the shaft, 1,600 feet (487m) below ground, and at the end of a long working day they had to climb up again.

GUNPOWDER AND FIREDAMP

The main improvement to mining came in the 17th century with the introduction of gunpowder for blasting. It still involved a great deal of hard work, as holes had to be drilled into the solid rock by hand to take the charge of powder, and it brought its own dangers in the days before the invention of the safety fuse. If the metal rod used to cram in the powder hit a rock and caused a spark, the whole charge could go off.

⌃ The pit bottom at Lady Victoria Colliery in Scotland; when they were first introduced the railed tracks made moving the coal far simpler.

One particular record described the terrible fate of a miner when an explosion fired the metal rod straight through his head. There was another danger that was emerging in coalmines. Dr Keys, the founder of Caius College, Cambridge, wrote in the mid 16th century: 'We also have in the northern part of Britain certain coal pits, the unwholesome vapour whereof is so pernicious to the hired labourers, that it would immediately destroy them, if they did not get out of the way as soon as their lamps became blue, and is consumed.' What they called firedamp we know as methane, and it was to become more of a problem as coalmines went ever deeper.

By the beginning of the 18th century, many mines had reached a depth where the available water-powered pumps were no longer able to cope. One man who was aware of the problem was Thomas Newcomen, a Dartmouth merchant who supplied iron tools to local miners. Mine pumps consisted of stout vertical timbers, the pump rods, attached to a piston that moved up and down in a cylinder, a sort of giant-sized version of the familiar village pump. Gravity took care of the down stroke: what was needed was power to raise the rods. In Newcomen's engine, the pump rods were suspended from one end of a beam, pivoted at its centre. At the other end of the beam, chains were connected to a piston that could move in a cylinder. The weight of rods at the opposite end of the beam lifted the piston. Steam from a boiler was then passed into the cylinder, which was then sprayed with cold water. The steam condensed, creating a vacuum, and air pressure on top of the piston forced it down. Pressure equalized, the whole cycle could start again and the pump would work. The engine was first installed at a colliery at Dudley, Staffordshire, in 1712, and was soon in use in mining districts throughout Britain.

Diagram of the Newcomen engine, installed to pump out water from the colliery at Dudley, in 1712.

BOULTON AND WATT

The Newcomen engine did the job, but it was very inefficient. Constantly cooling and reheating the cylinder used a huge amount of energy. This was not a problem at collieries, where the one thing that was abundantly available was coal, but it was a different matter in Devon and Cornwall, where coal had to be brought in. One mine agent in Cornwall complained that the mine might have to close because of the cost of coal 'which sweeps away all the profit'. A solution was found when a model of a Newcomen engine was sent to the instrument maker at Glasgow University, James Watt. He recognized the problem and built a new type of engine in which the steam was condensed in a separate vessel. This meant that the steam cylinder could be kept permanently hot. But heat

A modern steam winding engine.

The steam-engine man was responsible for the safety of the men in the shaft: he kept a keen eye on the dials that indicated the position of the miners' cage in the shaft to ensure it always came smoothly to a halt.

still escaped through the open top. Watt then closed the cylinder, and instead of relying on atmospheric pressure to push down the piston, he used steam pressure. He went into partnership with a Birmingham manufacturer, Matthew Boulton, and between 1777 and the end of the century they supplied more than 50 engines to Cornish mines. These were huge machines, with the overhead beam supported by one of the walls of the engine house. Few of the engines survive, but the engine houses have become a familiar part of the Cornish landscape.

TREVITHICK TO STEPHENSON

The firm of Boulton & Watt charged a premium for the use of the engines, which was calculated as one third of the saving between using their engine and an equivalent Newcomen engine. It was a good deal for the Cornish, but it did not stop Cornish engineers trying to make their own improvements, in spite of the Watt patent that was supposed to stop them. As soon as Watt's patent expired in 1800 there was a rush to develop new ideas. The Cornish mine engineer Richard Trevithick worked with high-pressure

steam that enabled him to get more power from a smaller engine – and he was able to make an engine that would move itself. His steam carriage received its first trial in Camborne in 1801, and within three years it had been developed into the first railway locomotive. Among those who developed his ideas was the engine man from Killingworth Colliery in Northumberland: George Stephenson.

Coal mines pioneered the development of the steam railway: here an imposing array of railway wagons have been loaded with coal.

AT THE COALFACE

At the end of the 18th century, the most common way of working a coal mine was a system known as 'pillar and stall'. The miners would only remove part of the seam, leaving the rest standing as supports. It was a wasteful method of working, as sometimes as much as half the coal had to be left in place to prevent the roof collapsing. Later the longwall method came into use, in which all the coal was removed. In the most efficient version, longwall retreating, headings (narrow passageways) were driven to the furthest point from the shaft, and the miners could start working across the seam at right angles to the heading. They worked back towards the shaft and, as the coalface moved, the supports were removed and the roof allowed to collapse behind the workers. But whichever method was used, the work was just as demanding.

CONFINED SPACES

In pillar and stall working, the stall was the area worked by one team, generally about 3 yards (3m) wide. The actual cutting of the coal was the work of the hewer, with the pickaxe as his principal tool. The first stage was to cut at the bottom of the seam to undermine the coal. Robert Bald, who wrote a book about Scottish mining in 1812, described the 'constant exertion and twisting of the body' that was required: 'For instance, it is a common practice for a collier, when making a horizontal cut in that part of the coal which is on a level with his feet, to sit down and place his right shoulder upon the right inside of his right knee; in this posture he will work long, and with good effect. At other times, he works sitting with his body half inclined to the one side, or stretched out his whole length, in seams of coal not thirty inches [70cm] thick.' It is almost impossible to imagine wielding a pickaxe while adopting what sounds like a yoga position. Once undercut, coal could be levered down.

➤ An 18th-century Glasgow pitman, taking his pickaxe for a day's work hewing at the coalface.

◄ Shovelling coal at the coalface in the early 20th century, a scene that scarcely changed throughout mining history.

THE NARROW SEAM

A Nottinghamshire miner remembered working in a narrow seam, measuring just 2 feet 10 inches (85cm) high: 'My father was 6 feet [1.8m] tall, weighed 15 stone [95kg], took a size 18 collar. When I first went down, I thought he'd never get in the seam, but he just threw his belly up and climbed in after it.'

⌃ Although modern powered supports replaced the old pit props, moving about in a narrow seam in the 20th century was no easier than it had been a century before.

near darkness it was often impossible to make fine distinctions. No account was taken of the fact that coal might have been spilled on its way from the face to the surface. A man could work all day and get nothing for his efforts, though the owners had coal to sell at a profit. It is perhaps not surprising that the history of mining is also the story of a long struggle between masters and men.

⌃ A low seam in the coalface: it is difficult to realize the effort involved in shovelling from a kneeling position.

When the coal had been broken down, it became the job of the putter to shovel it into the trucks, generally known as corves. It might seem that this was far easier work, but the putter would be working in the same narrow seam. Often he had to shovel from a kneeling position, throwing the coal above shoulder height to get it into the corve, with all the strain coming on his arms and shoulders.

EARNING A PITTANCE

The men were paid according to the number of corves they sent to the top, which was a source of a great many bitter disputes. If a corve was not completely filled, the miners did not lose just a part of its value – they lost it all. If it contained stone as well as coal, they could suffer the same fate. The men were working in those early days by the light of candles, and in the

⌃ After a morning's hard labour, the meal break had to be taken below ground: this miner is probably quenching his thirst with cold tea.

Women were employed underground in many mines until well into the 19th century. They all worked hard, but none endured such extremes of suffering as the women in Scottish collieries. The historian Robert Bald gave a very full description of their lives. Many of the young women had families. Their day started at dawn, when they bundled their infants into shawls to take them to baby-minders, before setting off for the pit. At the coalface, the men loaded the coal into wicker baskets, which were hoisted onto the women's backs. Bald measured the loads at one colliery and found that the average was 170lbs (77kg). The women carried them around 150 yards (137m) up a steep slope to the foot of the shaft. They then climbed 177-foot (53-m) long ladders to the surface, followed by a final short trudge to the coal store, where they could empty their loads. He found some women made the journey as many as 24 times a day. During that time they would have carried almost two tons of coal and climbed a height equivalent to that of a Scottish mountain.

➤ Scottish women carrying coals up ladders from the pit bottom to the surface.

ACCIDENTS

Records of accidents – both fatal and non-fatal – that befell women in Scottish mines make distressing reading. In 1890 Margaret Bissett, a 56-year-old trimmer at the Townhill pit in Fife, had a leg amputated after being run over by a wagon. The report concluded: 'Not yet resumed work.'

A WOMAN'S WORK IS NEVER DONE

Not surprisingly, Bald found many women 'weeping most bitterly from the excessive severity of the labour'. Their work down the pit was only a part of their lives. They had to go home in their pit clothes, covered in coal dust and often soaking wet. There they had to set to and light a fire to prepare meals for the rest of the family. It was impossible to maintain proper standards of care, and in many mining communities the infant death rate was almost as high as the birth rate. For all this effort and misery, the women were paid 8d (3p) a day. The Mines Act of 1842 finally made it illegal to employ women underground.

BAL MAIDENS

Women also played an important role working above ground, especially in Cornwall, where they were employed to break up the ore as it was brought to the surface. They were known as 'bal maidens' and there were two parts to the work. 'Spalding' involved breaking up the biggest rocks with long-handled hammers. The next stage was

◄ Bal maidens breaking up ore at the Carn Brea mine in Cornwall.

'SORE, SORE WORK'

Robert Bald described meeting one Scottish woman, 'groaning under an excessive weight of coals, trembling in every nerve, and almost impossible to keep her knees from sinking under her. On coming up, she said in a most plaintive and melancholy voice: "O Sir, this is sore, sore work. I wish to God that the first woman who tried to bear coals had broke her back, and none would have tried it again."'

▲ An illustration of a female coal-bearer: such women would have carried loads of around 170lbs (77kg).

'bucking', breaking down the larger rocks on an anvil. The females worked in sheds that were entirely open along one side, and as many mines were set high up on windswept moors or on the tops of cliffs, winter working was cold and miserable. They usually wore distinctive headdresses, popularly known as 'a yard of cardboard'. It was no more than a band of card passed round the head, with a length of material attached to act as a simple sunbonnet. One famous Cornish woman did work below ground with the men: Gracey Briney dressed like a man and on paydays went off with them to drink and smoke in the local pub.

CHILDREN IN THE MINES

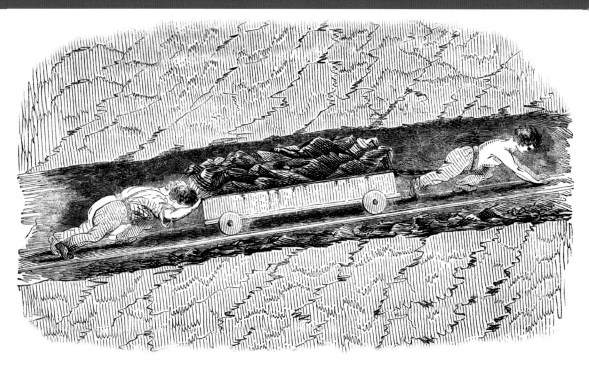

∧ Children working together to move a loaded wagon.

The world at large knew virtually nothing about children working down mines until a government report of 1842 revealed the full horror of their lives. The youngest were the trapper boys, whose job was to open and close the underground doors that controlled the flow of air through the workings. This was vitally important work, yet it was entrusted to children as young as five, who would sit at their posts for long hours, often in total darkness. One seven-year-old said he had been down the pit for three years: 'When I first went down I couldn't keep my eyes open. I don't fall asleep now, I smoke my pipe.'

A young trapper boy called Cooper left his post at Wellington Pit, Northumberland, on 19 April 1841. As a result there was an explosion in which 32 lives were lost, and the public enquiry blamed it on Cooper's 'gross negligence'. No one seems to have questioned the wisdom of trusting the safety of the whole mine to a seven-year-old boy.

HAULING COAL

When children grew bigger and stronger, they were employed as hauliers, moving the coal from the face to the bottom of the shaft. Janet Cummings, just 11 years old, went down the mine at five in the morning and came up again 12 hours later. She carried big lumps of coal in her hand, or smaller pieces in a basket, weighing up to a hundredweight (50kg), and said, 'The roof is very

∧ Children were harnessed to the coal trucks: this illustration is from the 1842 government report on the work of children in mines.

POOR WILLIAM

William Forrest was a pauper child, sent to the mine by the local poor house. He was only eight years old when he was first harnessed to a cart. He worked in water that was knee deep, and the roof was so low that he had to move along balancing on hands and feet. As a pauper, with no family to protect him, he was cruelly treated, beaten with a strap whenever he stopped work.

▲ The introduction of pit ponies in the mid-18th century did a great deal to ease the labour of children, who often became very fond of the animals they worked with.

low and I have to bend my back and legs and the water comes frequently up to the calves of my legs.' She had to carry her loads for a quarter of a mile (0.4km) in these conditions.

The use of trucks running on rails did not necessarily make the work any easier. At a colliery in Northumberland, boys and girls worked in a 30-inch (0.75-m) seam, and pushed the trucks along with their heads. They took their stockings off, wrapped them in their caps and used them as cushions for their heads, but even so one girl was able to demonstrate where she had been left with a bald patch. In other mines, the children wore harnesses and pulled the trucks along.

The introduction of pit ponies in some mines made for an easier life for the boys in charge of them, but they always had to keep control of the ponies and their loads. As one youngster reported:

'Sometimes [the boy] has to stop the tram suddenly. In an instant he is between the rail and the side of the level and in almost total darkness, slips a sprig between the spoke of his tram wheel and is back in his place with amazing dexterity, though it must be confessed, with all his activity, he is frequently crushed.'

SHOCKING IMMORALITY

Public opinion was less horrified by the working conditions than by what was seen as the shocking immorality of girls and boys working together, often naked from the waist up. Some teenage girls had to work with men who were stark naked because of the extreme heat. Most of the mine owners defended the conditions, declaring that the children were 'brought up to working habits' and denying that 'there is anything in the nature of employment in coal and iron mines that affect the health'. Parliament disagreed. The Mines Act of 1842 made it illegal to employ anyone under the age of ten in mines: it was a start.

▲ This illustration from the 1842 government report shocked society because the boy and girl were together and half naked.

COLEG RHYL COLLEGE 203394
CANOLFAN ADNODDAU LLYFRGELL
LIBRARY RESOURCE CENTRE

THE SAFETY LAMP

The greatest danger in coal mines was the presence of methane gas. For years the only answer relied on one brave miner who, when gas was detected, would put on a hooded robe of wet sacking and crawl up to the leak. The person who this task fell to was known as 'The Penitent' because his robes resembled those of someone serving a penance. When he reached the gas, he would light a candle stuck on the end of a long stick and poke it forward to ignite the methane. In theory the flames from the resulting explosion would pass over his head and he could make his escape. It was not satisfactory, and as mines were sunk ever deeper so dangers increased. The inspector of mines for Northumberland and Durham listed mining accidents between 1790 and 1840. Just over half of the 147 accidents were due to explosions, but they accounted for 1,243 of the 1,468 deaths. One of these explosions had long-lasting effects.

Before the introduction of the safety lamp, pockets of gas were ignited by 'The Penitent' using a candle at the end of a long stick: the sacking robes were supposed to protect him from the flames.

TRIGGERING AN INVENTION

The accident happened at Felling Colliery on Tyneside on 25 May 1812. The explosion was so fierce that it was heard up to 4 miles (6.5km) away and when all rescue attempts had ended it was found that 92 miners had died, the youngest of them a boy of ten. A local clergyman, Rev John Hodgson, wrote a full account of the disaster in the local paper. He managed to rouse public opinion and a Society for the Prevention of Accidents in Coal Mines was formed. It was well known that the main problem was the lack of any suitable lighting system in mines, apart from candles with naked flames, although various other devices had been tried – including a steel mill that sparked and the luminescence from rotting fish. And so it was decided to approach one of the leading scientists of the day, Sir Humphry Davy, to ask if he would consider trying to devise some form of lamp that could be used with safety.

SIR HUMPHRY DAVY (1778–1829)

Humphry Davy was born in Penzance, Cornwall, the son of a woodcarver. He left school at 15 to take a job as apprentice to a local apothecary. He continued with a programme of self-education and was fascinated by the new discoveries in science being made by men such as the French chemist Antoine Lavoisier. Davy was fortunate to attract the attention of the local MP Davies Giddy, who took him under his wing and helped him to get a place at Oxford University. By 1801 Davy had acquired a job at the Royal Institution, London, where he proved to be a brilliant lecturer and eventually became Professor of Chemistry.

DAVY'S DISCOVERY

Davy visited the north-east to investigate the problem at first hand, and in November 1815 he presented the results of his experiments. He discovered that if the naked flame was surrounded by a fine-mesh wire gauze, it could be safely used without causing explosions. Another benefit soon appeared. If gas was present, the flame in the lamp turned blue, so it could be used to test for the presence of methane. It became standard practice for mine deputies to take a lamp below ground before the start of a new shift to give their approval for work to start.

⋀ A group of miners with their safety lamps.

⋀ Sir Humphry Davy demonstrating his newly invented safety lamp.

⋀ The lamp room in a 20th-century colliery: although electric helmet lights were now in use, the Davy lamp on the table was still used to test for gas.

Davy was not the only inventor who produced a safety lamp: that same year George Stephenson produced his own version, which became known as the Geordie lamp and was used for many years in the north-east.

It was thought that the invention of the safety lamp would prevent serious accidents: it failed to do so. There were many reasons, including the fact that the mine owners did not supply lamps and the miners could not afford them. One of the main causes of explosions was poor ventilation. Darley Main Colliery, at Barnsley in South Yorkshire, was the scene of a whole series of accidents. On 14 April 1843 an explosion killed one man and injured another. In February 1847, when an explosive charge was let off it started a fire and although there was a valiant attempt to put out the flames, the men had to give up: six collapsed before reaching safety and died. Just two months later another explosion killed two men, and the coroner's jury added a rider that 'through the numerous accidents in this pit there must be some neglect on the part of the managers'. But nothing was done, and men still used naked candles for light. The worst disaster of them all occurred in 1849, when a huge explosion killed 75 miners. Once again the jury was critical of the management. However, no action was taken.

➤ The Fife mine rescue team photographed in the early years of the 20th century; the men have crude breathing apparatus.

⋏ The 1866 explosion at Oaks Colliery, Barnsley, in which 361 miners died.

Each miner going below ground is given a numbered tag so that the authorities know how many men are in the mine in the event of an accident.

Attempts by Parliament to enforce safety rules by appointing inspectors did very little to help, largely because there were too few to be effective. The inspector for South Wales, working full time, took three and a half years to visit every mine on his books. The laws were rarely enforced. Eleven members of a coroner's jury found an overseer guilty of gross negligence and manslaughter; the twelfth, himself an overseer, disagreed, and the coroner took the view of the one against the eleven. An Act of 1855 ruled that an owner found guilty of negligence could be fined; a miner found guilty of the same offence would go to gaol. Not surprisingly, the miners were unimpressed.

FRIENDS IN NEED

A feature of all mine disasters was the unstinting work of the miners themselves to do all they could to save the lives of their comrades. The rescue operation at Tynewydd Pit in South Wales in 1877 was typical. The men had inadvertently broken into old flooded workings and water came rushing in. Four men died instantly, swept away and drowned, while others were brought safely out. Five men were left, trapped in an air pocket, and heroic efforts were made to reach them. Divers volunteered, but failed to reach the men; the only way to get to them was to hack through a great wall of coal. An account described the men 'beating against the black face of coal, which at any moment might open out

From 1911, canaries were used to test for gas in mines because they reacted to the toxic fumes before humans. This one is in a humane cage, which has an air supply to revive an affected bird. Use of canaries was phased out in 1986, when electronic detectors were introduced.

and destroy them, they never turned their heads. With blood streaming … from their hands, yet they rained blow after blow and, said a looker on, never turned or paused'. Fortunately, their heroic efforts succeeded.

OTHER DANGERS

What was not understood in the 19th century was that methane was not the only danger in the mines. The introduction of more efficient machinery created fine clouds of coal dust that were also explosive. It was a coal dust explosion that led to Britain's worst-ever mine disaster at Senghenydd, Caerphilly, South Wales, in October 1913, in which 483 miners were killed.

The mines of the south-west of England were free from gas, but were prone to flooding. In 1846 Wheal Rose, near St Newlyn East in Cornwall, was suddenly inundated, killing 39 miners. Another Cornish disaster happened in 1919 when part of a steam engine crashed down the shaft at Levant mine, killing 31 and injuring many more.

Mining in Devon and Cornwall was organized in a unique way. The work of a mine was put up for auction at regular intervals. Men formed gangs and bid for particular jobs and sections. There were three distinct types of work: tutwork was paid by the measure, for example for sinking a shaft to a certain depth or opening up a new section of the mine; tribute was paid for winning the ore, and the amount the men got depended on the value of the ore they produced in a set period of time; dressing was separating the ore from the waste rock, and was generally taken by a man who employed a team of bal maidens.

TUTWORK

Tutwork was usually offered up first and the team who made the lowest bid got the job. They were known as a 'pair', no matter how many were in the group. They generally split into corps of two or three, who worked in shifts, day and night. The mining company provided them with candles, gunpowder and any special equipment they might need, but the cost was taken off their final payment. They also got an advance of cash, known as a subsist, so that when payday finally

▲ Miners preparing to go down the Dolcoath Mine in Cornwall in the 1890s.

arrived, by the time all the costs and subsist had been deducted there was often nothing much left, and the men had no option but to bid to go back to work again.

TRIBUTE

Tribute was more complex. The men had to use their own judgement to work out what they thought the ore might be worth. If the lode they were working proved to be very rich, they would often work up to 16 hours a day to make as much as they could in their set time. If, however, the ore suddenly petered out they could find themselves badly off. One group at Devon Great Consols mine worked for two months, at the end of which

▲ Loading a truck from a kibble in a Cornish mine.

> South Crofty mine, Cornwall, in 1871. Engine houses are still a familiar feature of the Cornish landscape.

they had just over £4 to share between them. The system had its faults, but it did mean that everyone who worked in a mine had a personal interest in making it a success.

The men working tribute had to follow a rich lode, no matter how far from the shaft they had to work. In places the ventilation was so bad that a boy had to be employed to waft air at the candles to keep them alight: if he went to sleep everyone would be plunged into Stygian gloom.

Conditions varied enormously: one miner reported how he often had to wade through icy cold water that reached up to his chest, while at a mine in the St Day area the temperature underground could reach 125°F (58°C). Some of the worst conditions were met in mines such as Botallack, where the workings stretched out under the sea. Here the men were often drenched with seawater and the salt chafed their limbs until they were red raw.

Largely because of these systems, the Cornish miners became famous for their expertise, and when work petered out in Britain many found jobs in mines all round the world.

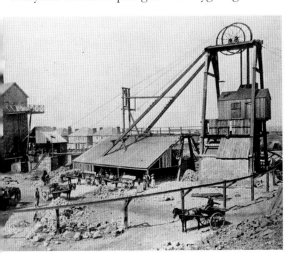

> East Pool mine, Cornwall, in 1895. Wires from the winding or whim engine on the left go to the headstock above the shaft. The engine is now preserved by the National Trust at this World Heritage Site.

LUNG DISEASE

A report of 1842 described the experiences of a 17-year-old working at Fowey: 'The air was "poor" where he then was and he had a pain in his head after working some time, which lasted for hours after he came to the surface. Almost every morning he had a cough and brought up some stuff black as ink.' The death rate from lung disease was even higher in the metal mines than in the collieries.

Ups and Downs

As mines went ever deeper, it became less and less practical for miners to clamber up and down ladders between the surface and the workings. The simplest alternative was to use a hand-operated winch, similar to those used for lowering buckets down wells. The miners either put their legs through loops or sat on a wooden rung attached to the rope. A more sophisticated answer was the horse gin. The haulage rope was wound round a drum mounted on a vertical post that could be rotated by a horse trudging round a circular track. These unfortunate beasts were described in a report on Yorkshire mines as 'of the worst description, spavined and blind'.

In coalmines the gins were mainly used for raising and lowering the kibbles, containers rather like huge buckets, that brought coal to the surface. Most mines used a balance system, in which an

⬆ A horse gin: the horse walking round a circular track turned the winding-drum that carried the cable for raising and lowering material in the shaft.

empty corve was lowered as the full one was raised. The corves were also used to carry miners, but accidents could easily be caused by the corves swaying and bumping in their passage up and down the shaft. In addition there was a danger from coal being dislodged. Eleven-year-old Enoch Hurst was hit on the head by falling coal and died.

▽ Near Broseley in Shropshire: a horse gin is in the background of this typical 18th-century pithead scene.

PIT PONIES

Many pit ponies spent most of their lives underground, working an eight-hour shift and rarely seeing the light of day. Getting the ponies underground in the first place was a difficult task, as was eventually bringing them up again, and involved much pushing and pulling to get them in the cage. As one miner reported, when a pony was released into an open field 'it was proper bedlam. They'd run about kicking and squealing and they seemed to go mad at first'.

> Pit ponies were well cared for in their underground stables.

The official report declared that no one was responsible and that the accident was his own fault: he should have got out of the way.

WINDING BY STEAM

The introduction of steam winding engines was in many ways a huge advance over the old horse gins, but brought their own dangers. In the early years the engine man had to keep a lookout for a piece of tallow tied to the rope. When that appeared in view, he knew that the kibble was near the top of the gantry. If he did not keep a careful watch the results could be disastrous, as a mine inspector explained: 'If on the contrary his attention is directed for a moment to another object, you are sent over the pulley with fearful rapidity and killed.' Later, more sophisticated indicator systems were introduced. Another danger was that the new steam winders put great strain on the ropes. One inspector was waiting to go down a pit when the rope hauling the kibble in which he was going to ride snapped: if it had broken a minute later he would have plunged to his death. The 1842 report on children in mines found that of 50 children who died in Yorkshire mines, 34 had been killed by accidents in shafts.

By the 1840s wire ropes had been introduced to replace hemp. Although they were far safer, some miners objected to them and even brought a court case to prevent them being used. The case was lost when the men could find no better objection other than they were new. The use of wire made winding much faster and it could take heavier loads. As a result, kibbles were gradually replaced by the cages running in guide rails that are still used today. They were made even safer by an invention of Edward Fourdrinier of Leek, Staffordshire. If a wire snapped, strong springs forced iron wedges against the sides of the shaft, bringing the cage to a halt.

⋀ Miners in the cage at the pit bottom.

Relations were always bad in the coal industry. A major source of conflict in the north-east of England was the Yearly Bond. Each year the miners had to sign up to agree to work for one particular colliery for the following year, in exchange for which they got a small signing-on fee. It was a very one-sided arrangement. If a miner took work of any sort anywhere else he was liable to be sent to gaol for breach of contract. The owners, however, were under no obligation to keep the miner in work: if demand was slack a miner could be laid off with no pay whatsoever. The Combination Acts of 1799 and 1800 decreed it illegal for workers to get together to demand their rights, and some owners took advantage of the situation to try and change the rules. They decided to alter the date of setting the Bond from October to January. In October the demand for coal was high so the men had a good bargaining position, but in January work was slack. Although unions were banned, a strike brought all the pits in the area to a standstill in November 1809. Rumours spread that the men were taking oaths of allegiance which justified the authorities' decision to bring in the military to break up any

It was common for men to work in gangs led by a 'butty', who received the pay on behalf of them all. Here colliers are sharing out the money.

protesting groups and arrest the ringleaders. The gaols were filled and the Bishop of Durham agreed to allow miners to be locked up in his stables. The mine owners took even more draconian measures. Most of the miners lived in houses rented from the companies; they were evicted with their wives and children and thrown into the street with all their possessions. With no funds to support them, the miners were eventually forced to submit. The Yearly Bond remained in place until 1872.

FURTHER CAUSE FOR CONFLICT

Another source of resentment was the system of payment by truck or tommy note. In the truck system, payment was made in goods instead of cash; with tommy they were paid with notes that could only be exchanged in the company's own stores. The system was defended by the owners who in 1796 claimed, 'The collier is not able to squander his gains, to the injury of himself and his family.' It was open to obvious abuse. In the 1820s, mine owners of Staffordshire and Shropshire attempted to reduce wages by 6d (2.5p) a day. This resulted in riots in which soldiers were called in: two miners were killed. Eventually an agreement was reached

Most miners rented houses from the company. When the men at Kinsley Colliery, Yorkshire, went on strike in 1905 they were evicted, and the children had to be fed in the street.

A LORD'S VIEW

In the 1920s, Lord Birkenhead, who had been involved on the government's behalf in trying to resolve various disputes in the coal industry, gave his personal verdict on the mining union leaders: 'I should call them the stupidest men in England if I had not previously had to deal with the owners.'

to only knock 4d (2p) off the pay, at which point the owners promptly raised the price of goods in the tommy shops.

Throughout the 19th and 20th centuries there was conflict between miners and owners. The repeal of the Combination Acts in the 1820s enabled the miners to form unions, but did little to prevent often-violent confrontations. The most famous action was the general strike of 1926, in which for a short time other unions joined the miners and brought many industries to a standstill.

➤ Miners, carrying a banner with portraits of early union leaders, march through the streets of Durham as they celebrate the annual Gala Day.

⌃ The organizers of a soup kitchen for striking miners in the 1920s at Newton, Midlothian.

In coal-mining districts, conditions during the 19th century were often poor. One Durham vicar described his village as a 'dark, dreary, dirty, smoky hole'. He commented that the houses were 'well ventilated', which turned out to mean that the front door opened directly into the kitchen/living room, and the air had nothing to impede it before it blew straight out the back door. Pithead baths were a rarity and the miner had to remove the dirt of the colliery by washing in a tin tub in the kitchen, with his wife usually being given the job of scrubbing his back. Sanitation was little better, with many terraces facing unpaved streets with open sewers. There were exceptions, however. Some colliery owners provided excellent accommodation. Earl Fitzwilliam owned mines near Rotherham in South Yorkshire, and some of the homes he provided still survive at Reform Row, Elsecar. They are solid, stone-built houses, with small front gardens and yards at the back.

Things were very different in the south-west of England. A Parliamentary Committee of 1864 reported that Camborne in Cornwall had over

Few miners' homes had the luxury of bathrooms in the early 20th century, and with no pithead baths the only answer was a tin tub in the kitchen.

800 good houses built, not by the company but by the men themselves. 'Do you consider this an evidence of their prudent habits?' one of the commissioners asked, to which the inspector replied, 'Of their prudent habits – yes – and of their temperate, orderly, and good conduct.'

COUNTRY COTTAGES

In country areas the miners agreed with the landowners to clear an area of ground – often blowing up rocks with gunpowder and using the stone to build their cottages. But they only held the cottages for three generations, after which the landowners took them back. Unlike the orderly terraces of the towns, these were often quite poor and badly situated. One writer described a typical village in Cornwall in the middle of the 19th century: 'The miners invariably occupy the most exposed and worst built cottages … surrounded by cesspools, broken roads and pools of undrained rain. The village of Amal-Voer is like a cluster of cottages huddled together on the top of a hill and scarcely space between them for access. The bedrooms are rarely more than one in each house, and open to the ceiling.'

Adams Row, Newton, Scotland, in the early part of the 20th century. Single-storey terraces such as these were typical of many Scottish mining villages.

Miners in the north-east of England were famous for their love of fine furniture to brighten their homes. One 19th-century writer noted: 'A chest of mahogany drawers, and an eight-day clock, with a mahogany case, are the great object of their ambitions.' The clock was a badge of success, and the family that acquired one would invite the whole neighbourhood to celebrate and admire it.

◁ A miner's children meet him as he returns home from work, c.1910.

A SPIRIT OF COMMUNITY

In spite of what was often the most wretched accommodation, all mining communities had their own organizations for mutual support. In the English Midlands, for example, many had boot clubs, which were just what the name suggests. Everyone contributed a shilling (5p) a week, and could draw out money when they needed new boots. Weddings and funerals were events that everyone attended. At the latter, in many villages it was the custom to bake a funeral cake and take a piece to every family in the community: to miss anyone out was a social disgrace.

The conditions were made worse for the miners by the fact that the houses were often some distance from the workings. The Parliamentary report of 1842 described men who worked in temperatures of 96°F (36°C) and were then faced with a 3-mile (5-km) tramp home through dark winter evenings, and 'often reached home without a fire and had to creep to bed with no more nourishing food and drink than barley-bread or potatoes with cold water'.

▷ Miners' wives and children outside their homes in Smeaton, Fife. Instead of terraces, some Scottish mining villages had rows of tenements like these.

THE FAMOUS ST HILDA COLLIERY BAND
WORLDS CHAMPIONS
1912-1920-1921-1924

The colliery band was an important part of many mining communities and brass band competitions were keenly contested.

One anonymous lady, married to a mine owner, wrote an account of life in a mining village in the 1860s. She was concerned about drunkenness and decided that she would do something positive about it by providing an alternative pastime. She started a choir, which on the whole was very successful, but if the men did not like the music she had chosen on any practice night they simply walked out and either went home or, more often, went back to the pub.

MAKING MUSIC

The tradition of the male voice choir was strongest in South Wales. Mining began in the Rhondda Valley in 1865 and within 50 years the population of the area had risen from less than 1,000 to 150,000. Choirs from the region became famous. A Rhondda choir took first prize at the Chicago World Trade Fair Eisteddfod in 1893, and the Treorchy choir was invited to sing at Windsor Castle for Queen Victoria.

Music-making of many kinds was popular and played an important part in special celebrations. At Christmas, miners crowded into the centres of Sunderland and Durham to perform sword dances. Rather like morris dancers, the men wore white shirts bedecked with ribbons, were led by a 'captain' in a cocked hat and accompanied by a fool, known as the 'Betsy', who took round a collecting box. Other Christmas celebrations included processing 'King Coal' – a huge block of coal garlanded with evergreens – all round the community to the accompaniment of a local choir.

The first mention of an actual colliery band appears in 1809. This would have been a woodwind band: the now familiar brass band only appeared in the 1830s. Many colliery owners encouraged the formation of bands and a tradition of band competitions soon developed. The first was held at Belle Vue, Manchester, in 1853 and attracted an audience of 16,000. Soon all miners' gatherings were famous for their bands. It began with the first Northumberland Miners' Picnic in 1867, closely followed by the famous Durham Miners' Gala in which the bands march through the street behind colourful banners. The Gala is still held annually and, even though the last colliery in the area has closed, it continues to attract crowds of 100,000 and more.

The famous Grimethorpe Colliery Band from North Yorkshire, formed in 1917, heard that the pit was to close just five days before competing in the 1992 National Brass Band Championships. They went on to win the competition, and still play under the same name at venues all over the country.

⌃ Miners were noted for being fond of gambling, and many kept whippets for racing.

SPORTING TIMES

Sport – particularly football, rugby and cricket – brought a welcome relief from toiling below ground. It was said in Yorkshire and Lancashire that if either county needed a new player, all they had to do was whistle down a mineshaft.

⌃ The whites of these Yorkshire miners make a startling contrast with the sooty clothes they would be wearing when they went back to work.

OTHER PASTIMES

Not all the miners spent their leisure time in such innocent pursuits. There was always a great love of gambling, often based on such simple games as quoits, and pitch and toss; others combined gambling with their hobbies of keeping whippets or racing pigeons. Sport in the early days was often alarmingly violent. The 1842 report on children in mines described 12 separate fights staged in Lancashire one Christmas: 'They fight quite naked excepting their clogs. When one has the other down on the ground, he first endeavours to choke him, then he kicks him in the head with his clogs. Sometimes they are very seriously injured, that man you saw today with a piece out of his shoulder is a great fighter.'

A very different pastime could be found in the Scottish lead mining district of Wanlockhead. In 1756, 32 miners got together and raised a subscription to open a library 'for our mutual improvement'. It was such a success that a century later another subscription was raised to pay for a new purpose-built library. It is still there today.

British tin and copper mines all suffered a catastrophic decline in their fortunes during the 19th century. Cornwall's copper mines had enjoyed a boom time in the 1850s, until huge copper deposits were found in America and Spain. The Cornish found it increasingly difficult to compete, and areas such as Gwennap that had once employed a thousand men saw mines closing at a steady rate, until by the end of the 1870s not one was left open. It was estimated that the number of working miners in Cornwall as a whole dropped by a third during this period.

PASTURES NEW

The Cornish miners, however, did not just sit at home and bemoan their fate: they knew they had skills with a market value, and set off to work wherever in the world they were wanted. The Anglo-Mexican Mining Association was offering £15 a month – six times what they would earn in Cornwall. It seemed too good to be true – and it was. The conditions were atrocious and out of one party of 48 who signed up, only 18 survived. Others were more fortunate, and in the 1890s special trains were being run from Cornwall to

∧ A 'Cousin Jack': a Cornish miner photographed in a Californian gold field.

Southampton to transport the migrating workers. Known as 'Cousin Jacks', a name of uncertain origin, they spread around the world. In one family all nine sons left, and unfortunately four died overseas – one in America, one in New Zealand, another in Australia and the fourth in South Africa.

The decline in lead mining was even more precipitous: by the end of the 19th century there were fewer than 300 men at work in Derbyshire. However, the effects there were not so dramatic as in the south-west of England, thanks to the development of new industries such as engineering.

∧ A mechanical cutter being used to undercut the coalface at Canderigg Colliery, Scotland.

△ Miners moving a conveyor belt that will carry coal from the working face to the shaft.

BEVIN BOYS

During the Second World War there was a serious shortage of labour in the coal mines because so many men had joined the forces. In December 1943, Ernest Bevin, Minister for Labour and National Service, devised a scheme where each month ten numbers were put into a hat and two drawn out; conscripts with call-up papers that ended in those numbers were sent to collieries instead of the armed services, along with a number of volunteers. The 48,000 who were called up for mine service became known as Bevin Boys.

ADVANCES IN TECHNOLOGY

Things looked very different in the coal industry. Demand had never been higher and by the beginning of the 20th century the British industry was employing over a million men. But changes were on the way, though they only came slowly. Attempts at producing a mechanical coal cutter had begun as long ago as 1768; however, it was to be another century before the first really successful machine went into production. It was rather like a circular chain saw set at the end of a jib, powered by compressed air; it was claimed that it could undercut a coalface to a depth of 4½ feet (1.4m) and could work a 200-yard (182m) face in six hours. Even with this big advance in technology, only 1.5 per cent of coal was being cut with machines at the start of the 20th century. The next big advance was the introduction of conveyor belts to the Durham coalfield in 1902, to move the coal from the face to the shaft. It did not necessarily make life easier for the miners. Men still had to shovel the coal onto the belt, usually from a kneeling position, hour after hour. One change did benefit the men: the 20th century finally saw the introduction of pithead baths at the majority of collieries.

Mechanization continued throughout the 20th century, but the demand for coal began to fall. Clean Air Acts reduced the demand for household coal; North Sea gas replaced town gas; the steam engine became obsolete. What no one in the industry could have foreseen was that the Britain that had exported coal to the world would end up importing it. In 2011 the UK consumed 51.2 million tonnes of coal, mainly in power stations, of which 32.4 million tonnes was imported. Today only three deep pits remain open: Daw Mill, Warwickshire; Kellingley, Yorkshire; and Thoresby, Nottinghamshire. An era has come to an end.

△ A maintenance man working on the head of a mechanical cutter.

The mines, museums and heritage centres listed below all provide insights into the history of mining. Contact them or visit their websites for further information and details of opening dates/times.

Astley Green Colliery Museum, Higher Green Lane, Astley Green, Tyldesley, Manchester M29 7JB

Beamish: the Living Museum of the North, Beamish, County Durham DH9 0RG

Big Pit: National Coal Museum, Blaenafon, Torfaen, South Wales NP4 9XP

Black Country Living Museum, Tipton Road, Dudley, West Midlands DY1 4SQ

Blue Hills Tin Streams, Wheal Kitty, St Agnes, Cornwall TR5 0YW

Cefn Coed Colliery Museum, Neath Road, Creunant, South Wales SA10 8SN

Dolaucothi Gold Mines, Pumsaint, Llanwarda, Carmarthenshire, West Wales SA19 8US

East Pool Mine, Pool, near Redruth, Cornwall TR15 3ED

Geevor Tin Mine, Pendeen, Penzance, Cornwall TR19 7EW

Haig Pit Mining and Colliery Museum, Solway Road, Kells, Whitehaven, Cumbria CA28 9BG

Hopewell Colliery Museum, Lacinda Coalway, Coleford, Gloucestershire GL16 7EL

Killhope: The North of England Lead Mining Museum; near Cowshill, Upper Weardale, County Durham DL13 1AR

King Edward Mine Museum, Troon, Camborne, Cornwall TR14 9DP

Levant Mine and Beam Engine, Trewellard, Pendeen, near St Just, Cornwall TR19 7SX

Morwellham Quay World Heritage Site, Morwellham, Tavistock, Devon PL19 8JL

Museum of Lead Mining, Wanlockhead, By Biggar, Lanarkshire, Scotland ML12 6UT

The Pit Village at Beamish Museum.

National Coal Mining Museum for England, Caphouse Colliery, New Road, Overton, Wakefield, West Yorkshire WF4 4RH

National Mining Museum Scotland, Lady Victoria Colliery, Newtongrange, Midlothian EH22 4QN

Nenthead Mines Heritage Centre, Alston, Cumbria, CA9 3PD

Peak District Mining Museum, The Pavilion, Matlock Bath, Derbyshire DE4 3NR

Poldark Mine, Wendron, Helston, Cornwall TR13 0ER

The Silver Mountain Experience, Llwyernog, Ponterwyd, Aberystwyth, Wales SY23 3AB

South Wales Miners' Museum, Afan Forest Park, Cynonville, Port Talbot, Wales SA13 3HG

Sygun Copper Mine, Beddgelert, Gwynedd, Wales LL55 4NE

Tolgus Tin Streaming Mill, New Portreath Road, Redruth, Cornwall TR16 4HN

Woodhorn Museum and Northumberland Archives, Queen Elizabeth II Country Park, Ashington, Northumberland NE63 9YF

Information correct at time of going to press.